Privacy and Data Security

Data security is crucial for all small businesses. Customer and client information, payment information, personal files, bank account details - all of this information is often impossible replace if lost and dangerous in the hands of criminals. Data lost due to disasters such as a flood or fire is devastating, but losing it to hackers or a malware infection can have far greater consequences. How you handle and protect your data is central to the security of your business and the privacy expectations of customers, employees and partners.

Cyber Plan Action Items:

1. Conduct an inventory to help you answer the following questions:

- **What kind of data do you have in your business?**

A typical business will have all kinds of data, some of it more valuable and sensitive than others, but all data has value to someone. Your business data may include customer data such as account records, transaction accountability and financial information, contact and address information, purchasing history, buying habits and preferences, as well as employee information such as payroll files, direct payroll account bank information, Social Security numbers, home addresses and phone numbers, work and personal email addresses. It can also include proprietary and sensitive business information such as financial records, marketing plans, product designs, and state, local and federal tax information.

- **How is that data handled and protected?**

Security experts are fond of saying that data is most at risk when it's on the move. If all your business-related data resided on a single computer or server that is not connected to the Internet, and never left that computer, it would probably be very easy to protect.

But most businesses need data to be moved and used throughout the company. To be meaningful data must be accessed and used by employees, analyzed and researched for marketing purposes, used to contact customers, and even shared with key partners. Every time data moves, it can be exposed to different dangers.

As a small business owner, you should have a straightforward plan and policy – a set of guidelines, if you like – about how each type of data should be handled, validated and protected based on where it is traveling and who will be using it.

- **Who has access to that data and under what circumstances?**

Not every employee needs access to all of your information. Your marketing staff shouldn't need or be allowed to view employee payroll data and your administrative staff may not need access to all your customer information.

When you do an inventory of your data and you know exactly what data you have and where it's kept, it is important to then assign access rights to that data. Doing so simply means creating a list of the specific employees, partners or contractors who have access to specific data, under what circumstances, and how those access privileges will be managed and tracked.

Your business could have a variety of data, of varying value, including:

- Customer sales records
- Customer credit card transactions
- Customer mailing and email lists
- Customer support information

- Customer warranty information
- Patient health or medical records
- Employee payroll records
- Employee email lists
- Employee health and medical records
- Business and personal financial records
- Marketing plans
- Business leads and enquiries
- Product design and development plans
- Legal, tax and financial correspondence

2. Once you've identified your data, keep a record of its location and move it to more appropriate locations as needed.

3. Develop a privacy policy

Privacy is important for your business and your customers. Continued trust in your business practices, products and secure handling of your clients' unique information impacts your profitability. Your privacy policy is a pledge to your customers that you will use and protect their information in ways that they expect and that adhere to your legal obligations.

Your policy starts with a simple and clear statement describing the information you collect about your customers (physical addresses, email addresses, browsing history, etc), and what you do with it. Customers, your employees and even the business owners increasingly expect you to make their privacy a priority. There are also a growing number of regulations protecting customer and employee privacy and often costly penalties for privacy breaches. You will be held accountable for what you claim and offer in your policy.

That's why it's important to create your privacy policy with care and post it clearly on your website. It's also important to share your privacy policies, rules and expectations with all employees and partners who may come into contact with that information. Your employees need to be familiar with your legally required privacy policy and what it means for their daily work routines.

Your privacy policy will should address the following types of data:

- **Personally Identifiable Information:** Often referred to as PII, this information includes such things as first and last names, home or business addresses, email addresses, credit card and bank account numbers, taxpayer identification numbers, patient numbers and Social Security numbers. It can also include gender, age and date of birth, city of birth or residence, driver's license number, home and cell phone numbers.

- **Personal Health Information**: Whether you're a healthcare provider with lots of sensitive patient information or you simply manage health or medical information for a small number of employees, it's vital that you protect that information. A number of studies have found most consumers are very concerned about the privacy and protection of their medical records. They do not want their health information falling into the hands of hackers or identity thieves who might abuse it for financial gain. But they also may not want employees or co-workers prying into their personal health details. And they often don't want future employers or insurers finding out about any medical conditions or history.

- **Customer information**: This includes payment information such as credit or debit card numbers and verification codes, billing and shipping addresses, email addresses, phone numbers, purchasing history, buying preferences and shopping behavior.

The Better Business Bureau has a copy of a privacy policy that you are free to download and use. It is available here: http://www.bbbonline.org/reliability/privacy/.

4. Protect data collected on the Internet

Your website can be a great place to collect information – from transactions and payments to purchasing and browsing history, and even newsletter signups, online enquiries and customer requests.

This data must be protected, whether you host your own website and therefore manage your own servers or your website and databases are hosted by a third party such as a web hosting company.

If you collect data through a website hosted by a third party, be sure that third party protects that data fully. Apart from applying all the other precautions that have been described, such as classifying data and controlling access, you need to make sure any data collected through your website and stored by the third party is sufficiently secure. That means protection from hackers and outsiders as well as employees of that hosting company.

5. Create layers of security

Protecting data, like any other security challenge, is about creating layers of protection. The idea of layering security is simple: You cannot and should not rely on just one security mechanism – such as a password – to protect something sensitive. If that security mechanism fails, you have nothing left to protect you.

When it comes to data security, there are a number of key procedural and technical layers you should consider:

Inventory your data

We mentioned before the need to conduct a data inventory so you have a complete picture of all the data your business possesses or controls. It's essential to get a complete inventory, so you don't overlook some sensitive data that could be exposed.

Identify and protect your sensitive and valuable data

Data classification is one of the most important steps in data security. Not all data is created equal, and few businesses have the time or resources to provide maximum protection to all their data. That's why it's important to classify your data based on how sensitive or valuable it is – so that you know what your most sensitive data is, where it is and how well it's protected.

Common data classifications include:

> HIGHLY CONFIDENTIAL: This classification applies to the most sensitive business information that is intended strictly for use within your company. Its unauthorized disclosure could seriously and adversely impact your company, business partners, vendors and/or customers in the short and long term. It could include credit-card transaction data, customer names and addresses, card magnetic stripe contents, passwords and PINs, employee payroll files, Social Security numbers, patient information (if you're a healthcare business) and similar data.

> SENSITIVE: This classification applies to sensitive business information that is intended for use within your company, and information that you would consider to be private should be included in this classification. Examples include employee performance evaluations, internal audit reports, various financial reports, product designs, partnership agreements, marketing plans and email marketing lists.

> INTERNAL USE ONLY: This classification applies to sensitive information that is generally accessible by a wide audience and is intended for use only within your company. While its unauthorized disclosure to

outsiders should be against policy and may be harmful, the unlawful disclosure of the information is not expected to impact your company, employees, business partners, vendors and the like.

Control access to your data

No matter what kind of data you have, you must control access to it. The more sensitive the data, the more restrictive the access. As a general rule, access to data should be on a need-to-know basis. Only individuals who have a specific need to access certain data should be allowed to do so.

Once you've classified your data, begin the process of assigning access privileges and rights – that means creating a list of who can access what data, under what circumstances, what they are and are not allowed to do with it and how they are required to protect it. As part of this process, a business should consider developing a straightforward plan and policy – a set of guidelines – about how each type of data should be handled and protected based on who needs access to it and the level of classification.

Secure your data

In addition to administrative safeguards that determine who has access to what data, technical safeguards are essential. The two primary safeguards for data are passwords and encryption.

Passwords implemented to protect your most sensitive data should be the strongest they can reasonably be. That means passwords that are random, complex and long (at least 10 characters), that are changed regularly and that are closely guarded by those who know them. Employee training on the basics of secure passwords and their importance is a must.

Passwords alone may not be sufficient to protect sensitive data. Businesses may want to consider two-factor authentication, which often combines a password with another verification method, such as a dynamic personal identification number, or PIN.

Some popular methods of two-factor identification include:
- Something the requestor individually knows as a secret, such as a password or a PIN.
- Something the requestor uniquely possesses, such as a passport, physical token or ID card.
- Something the requestor can uniquely provide as biometric data, such as a fingerprint or face geometry.

Another essential data protection technology is encryption. Encryption has been used to protect sensitive data and communications for decades, and today's encryption is very affordable, easy-to-use and highly effective in protecting data from prying eyes.

Encryption encodes or scrambles information to such an advanced degree that it is unreadable and unusable by anyone who does not have the proper key to unlock the data. The key is like a password, so it's very important that the key is properly protected at all times.

Encryption is affordable for even the smallest business, and some encryption software is free. You can use encryption to encrypt or protect an entire hard drive, a specific folder on a drive or just a single document. You can also use encryption to protect data on a USB or thumb drive and on any other removable media.

Because not all levels of encryption are created equal, businesses should consider using a data encryption method that is FIPS-certified (Federal Information Processing Standard), which means it has been certified for compliance with federal government security protocols.

Back up your data

Just as critical as protecting your data is backing it up. In the event that your data is stolen by thieves or hackers, or even erased accidentally by an employee, you will at least have a copy to fall back on.

Put a policy in place that specifies what data is backed up and how; how often it's backed up; who is responsible for creating backups; where and how the backups are stored; and who has access to those backups.

Small businesses have lots of affordable backup options, whether it's backing up to an external drive in your office, or backing up automatically and online so that all your data is stored at a remote and secure data center.

Remember, physical media such as a disc or drive used to store a data backup is vulnerable no matter where it is, so make sure you guard any backups stored in your office or off site and also make sure that your backup data storage systems are encrypted.

6. Plan for data loss or theft

Every business has to plan for the unexpected, and that includes the loss or theft of data from your business. Not only can the loss or theft of data hurt your business, brand and customer confidence, it can also expose you to the often-costly state and federal regulations that cover data protection and privacy. Data loss can also expose businesses to significant litigation risk.

That's why it's critical to understand exactly what data or security breach regulations affect your business and how prepared you are to respond to them. That should be the foundation of a data breach response plan that will make it easier to launch a rapid and coordinated response to any loss or theft of data.

At the very least, all employees and contractors should understand that they must immediately report any loss or theft of information to the appropriate company officer. And because data privacy and breach laws can be very broad and strict, no loss should be ignored. So even if you have sensitive data that just can't be accounted for, such as an employee who doesn't remember where he left a backup tape, it may still constitute a data breach and you should act accordingly.

And just in case you don't think a data breach could happen at your small business, think about this. In 2010, the U.S. Secret Service and Verizon Communications Inc.'s forensic analysis unit responded to a combined 761 data breaches. Of those, 482, or 63 percent, were at companies with 100 employees or fewer. And in 2011 Visa estimated that about 95 percent of the credit-card data breaches it discovers are on its smallest business customers.

The Online Trust Alliance has a comprehensive guide to understand and preparing for data breaches, available at https://otalliance.org/resources/2011DataBreachGuide.pdf.

The Federal Trade Commission has materials to help small businesses secure data in their care and protect their customers' privacy, including an interactive video tutorial, at http://business.ftc.gov/privacy-and-security.

Scams and Fraud

New telecommunication technologies may offer countless opportunities for small businesses, but they also offer cyber criminals many new ways to victimize your business, scam your customers and hurt your reputation. Businesses of all sizes should be aware of the most common scams perpetrated online.

To protect your business against online scams, be cautious when visiting web links or opening attachments from unknown senders, make sure to keep all software updated, and monitor credit cards for unauthorized activity.

Cyber Plan Action Items:

1. Train employees to recognize social engineering

Social engineering, also known as "pretexting," is used by many criminals, both online and off, to trick unsuspecting people into giving away their personal information and/or installing malicious software onto their computers, devices or networks. Social engineering is successful because the bad guys are doing their best to make their work look and sound legitimate, sometimes even helpful, which makes it easier to deceive users.

Most offline social engineering occurs over the telephone, but it frequently occurs online, as well. Information gathered from social networks or posted on websites can be enough to create a convincing ruse to trick your employees. For example, LinkedIn profiles, Facebook posts and Twitter messages can allow a criminal to assemble detailed dossiers on employees. Teaching people the risks involved in sharing personal or business details on the Internet can help you partner with your staff to prevent both personal and organizational losses.

Many criminals use social engineering tactics to get individuals to voluntarily install malicious computer software such as fake antivirus, thinking they are doing something that will help make them more secure. Fake antivirus is designed to steal information by mimicking legitimate security software. Users who are tricked into loading malicious programs on their computers may be providing remote control capabilities to an attacker, unwittingly installing software that can steal financial information or simply try to sell them fake security software. The malware can also make system modifications which make it difficult to terminate the program. The presence of pop-ups displaying unusual security warnings and asking for credit card or personal information is the most obvious method of identifying a fake antivirus infection.

2. Protect against online fraud

Online fraud takes on many guises that can impact everyone, including small businesses and their employees. It is helpful to maintain consistent and predictable online messaging when communicating with your customers to prevent others from impersonating your company.

Be sure to never request personal information or account details through email, social networking or other online messages. Let your customers know you will never request this kind of information through such channels and instruct them to contact you directly should they have any concerns.

3. Protect against phishing

Phishing is the technique used by online criminals to trick people into thinking they are dealing with a trusted website or other entity. Small businesses face this threat from two directions -- phishers may be impersonating them to take advantage of unsuspecting customers, and phishers may be trying to steal their employees' online credentials. Attackers often take advantage of current events and certain types of the year, such as:

- Natural disasters (Hurricane Katrina, Indonesian tsunami)

- Epidemics and health scares (H1N1)
- Economic concerns
- Major political elections
- Holidays

Businesses should ensure that their online communications never ask their customers to submit sensitive information via email, personal visits, or phone. Make a clear statement in your communications reinforcing that you will never ask for personal information via email so that if someone targets your customers, they may realize the request is a scam.

Employee awareness is your best defense against your users being tricked into handing over their usernames and passwords to cyber criminals. Explain to everyone that they should never respond to incoming messages requesting private information. If a stranger claims to be from a legitimate organization, verify his or her identity with his or her stated company before sharing any personal or classified information. Also, to avoid being led to a fake site, employees should know to never click on a link sent by email from an untrustworthy source. Employees needing to access a website link sent from a questionable source should open an Internet browser window and manually type in the site's web address to make sure the emailed link is not maliciously redirecting to a dangerous site.

This advice is especially critical for protecting online banking accounts belonging to your organization. Criminals are targeting small business banking accounts more than any other sector. If you believe you have revealed sensitive information about your organization, make sure to:

- Report it to appropriate people within your organization
- Contact your financial institution and close any accounts that may have been compromised (if you believe financial data is at risk)
- Change any passwords you may have revealed, and if you used the same password for multiple resources, make sure to change it for each account

4. Don't fall for fake antivirus offers

Fake antivirus, "scareware" and other rogue online security scams have been behind some of the most successful online frauds in recent times. Make sure your organization has a policy in place explaining what the procedure is if an employee's computer becomes infected by a virus.

Train your employees to recognize a legitimate warning message (using a test file from eicar.org, for example) and to properly notify your IT team if something bad or questionable has happened.
If possible, configure your computers to not allow regular users to have administrative access. This will minimize the risk of them installing malicious software and condition users that adding unauthorized software to work computers is against policy.

5. Protect against malware

Businesses can experience a compromise through the introduction of malicious software, or malware. Malware can make its way onto machines from the Internet, downloads, attachments, email, social media, and other platforms. One specific malware to be aware of is key logging, which is malware that tracks a user's keyboard strokes.

Many businesses are falling victim to key-logging malware being installed on computer systems in their environment. Once installed, the malware can record keystrokes made on a computer, allowing bad guys to see passwords, credit card numbers and other confidential data. Keeping security software up to date and patching your computers regularly will make it more difficult for this type of malware to infiltrate your network.

6. Develop a layered approach to guard against malicious software

Despite progress in creating more awareness of security threats on the Internet, malware authors are not giving up. The malware research firm SophosLabs reports seeing more than 100,000 unique malicious software samples every single day.

Effective protection against viruses, Trojans and other malicious software requires a layered approach to your defenses. Antivirus software is a must, but should not be a company's only line of defense. Instead, deploy a combination of many techniques to keep your environment safe.

Also, be careful with the use of thumb drives and other removable media. These media could have malicious software pre-installed that can infect your computer, so make sure you trust the source of the removable media devices before you use them.

Combining the use of web filtering, antivirus signature protection, proactive malware protection, firewalls, strong security policies and employee training significantly lowers the risk of infection. Keeping protection software up to date along with your operating system and applications increases the safety of your systems.

7. Be aware of spyware and adware

Spyware and adware, when installed will send pop-up ads, redirect to certain websites, and monitor websites that you visit. Extreme versions can track what keys are typed. Spyware can cause your computer to become slow and also leaves you susceptible to privacy theft. If you are subject to endless pop-up windows or are regularly redirected to websites other than what you type in your browser, your computer is likely infected with spyware.

To remove spyware run an immediate full scan of your computer with anti-virus software and if necessary run a legitimate product specifically designed to remove spyware. To avoid being infected with spyware, limit cookies on your browser preferences, never click on links within pop-up windows, and be wary of free downloadable software from unreputable sources.

8. Verify the identity of telephone information seekers

Most offline social engineering occurs over the telephone. Information gathered through social networks and information posted on websites can be enough to create a convincing ruse to trick your employees.
Ensure that you train employees to never disclose customer information, usernames, passwords or other sensitive details to incoming callers. When someone requests information, always contact the person back using a known phone number or email account to verify the identity and validity of the individual and their request.

Helpful links

- Use the Department of Homeland Security's Stop.Think.Connect.™ Campaign's resources created especially for businesses to train their employees: www.dhs.gov/stopthinkconnect
- Find the most updated patches for your computer and software applications: http://www.softwarepatch.com/
- Free computer security scan tools for your PC or network: http://www.staysafeonline.org/tools-resources/free-security-check-ups
- Stay on top of the latest scams, frauds and security threats as they happen: http://nakedsecurity.sophos.com/
- Additional tops to prevent against phishing: http://www.fraud.org/tips/internet/phishing.htm
- Learn how to resist phishing techniques with this interactive game: http://cups.cs.cmu.edu/antiphishing_phil/

Network Security

Securing your company's network consists of: (1) identifying all devices and connections on the network; (2) setting boundaries between your company's systems and others; and (3) enforcing controls to ensure that unauthorized access, misuse, or denial-of-service events can be thwarted or rapidly contained and recovered from if they do occur.

Cyber Plan Action Items:

1. Secure internal network and cloud services

Your company's network should be separated from the public Internet by strong user authentication mechanisms and policy enforcement systems such as firewalls and web filtering proxies. Additional monitoring and security solutions, such as anti-virus software and intrusion detection systems, should also be employed to identify and stop malicious code or unauthorized access attempts.

Internal network

After identifying the boundary points on your company's network, each boundary should be evaluated to determine what types of security controls are necessary and how they can be best deployed. Border routers should be configured to only route traffic to and from your company's public IP addresses, firewalls should be deployed to restrict traffic only to and from the minimum set of necessary services, and intrusion prevention systems should be configured to monitor for suspicious activity crossing your network perimeter. In order to prevent bottlenecks, all security systems you deploy to your company's network perimeter should be capable of handling the bandwidth that your carrier provides.

Cloud based services

Carefully consult your terms of service with all cloud service providers to ensure that your company's information and activities are protected with the same degree of security you would intend to provide on your own. Request security and auditing from your cloud service providers as applicable to your company's needs and concerns. Review and understand service level agreements, or SLAs, for system restoration and reconstitution time.

You should also inquire about additional services a cloud service can provide. These services may include backup-and-restore services and encryption services, which may be very attractive to small businesses.

2. Develop strong password policies

Generally speaking, two-factor authentication methods, which require two types of evidence that you are who you claim to be, are safer than using just static passwords for authentication. One common example is a personal security token that displays changing passcodes to be used in conjunction with an established password. However, two-factor systems may not always be possible or practical for your company.

Password policies should encourage your employees to employ the strongest passwords possible without creating the need or temptation to reuse passwords or write them down. That means passwords that are random, complex and long (at least 10 characters), that are changed regularly, and that are closely guarded by those who know them.

3. Secure and encrypt your company's Wi-Fi

Wireless access control

Your company may choose to operate a Wireless Local Area Network (WLAN) for the use of customers, guests and visitors. If so, it is important that such a WLAN be kept separate from the main company network so that traffic from the public network cannot traverse the company's internal systems at any point.

Internal, non-public WLAN access should be restricted to specific devices and specific users to the greatest extent possible while meeting your company's business needs. Where the internal WLAN has less stringent access controls than your company's wired network, dual connections -- where a device is able to connect to both the wireless and wired networks simultaneously -- should be prohibited by technical controls on each such capable device (e.g., BIOS-level LAN/WLAN switch settings). All users should be given unique credentials with preset expiration dates to use when accessing the internal WLAN.

Wireless encryption

Due to demonstrable security flaws known to exist in older forms of wireless encryption, your company's internal WLAN should only employ Wi-Fi Protected Access 2 (WPA2) encryption.

4. Encrypt sensitive company data

Encryption should be employed to protect any data that your company considers sensitive, in addition to meeting applicable regulatory requirements on information safeguarding. Different encryption schemes are appropriate under different circumstances. However, applications that comply with the OpenPGP standard, such as PGP and GnuPG, provide a wide range of options for securing data on disk as well as in transit. If you choose to offer secure transactions via your company's website, consult with your service provider about available options for an SSL certificate for your site.

5. Regularly update all applications

All systems and software, including networking equipment, should be updated in a timely fashion as patches and firmware upgrades become available. Use automatic updating services whenever possible, especially for security systems such as anti-malware applications, web filtering tools and intrusion prevention systems.

6. Set safe web browsing rules

Your company's internal network should only be able to access those services and resources on the Internet that are essential to the business and the needs of your employees. Use the safe browsing features included with modern web browsing software and a web proxy to ensure that malicious or unauthorized sites cannot be accessed from your internal network.

7. If remote access is enabled, make sure it is secure

If your company needs to provide remote access to your company's internal network over the Internet, one popular and secure option is to employ a secure Virtual Private Network (VPN) system accompanied by strong two-factor authentication, using either hardware or software tokens.

8. Create Safe-Use Flash Drive Policy

Ensure employees never put any unknown flash drive or USBs into their computer. As the U.S. Chamber's *Internet Security Essentials for Business 2.0* states, small businesses should set a policy so that employees know they should never open a file from a flash drive they are not familiar with and should hold down the Shift key when inserting the flash drive to block malware.

Helpful links

- Microsoft Password Strength Checker:
  https://www.microsoft.com/security/pc-security/password-checker.aspx
- Philip Zimmerman, Where to Get PGP:
  http://philzimmermann.com/EN/findpgp/
- US-CERT Security Publications:
  http://www.us-cert.gov/reading_room/
- NIST Special Publication 800-153, Draft Guidelines for Securing Wireless Local Area Networks (WLANs):
  http://csrc.nist.gov/publications/drafts/800-153/Draft-SP800-153.pdf
- U.S. Chamber of Commerce: Internet Security Essentials for Business 2.0
  https://www.uschamber.com/sites/default/files/issues/technology/files/ISEB-2.0-CyberSecurityGuide.pdf

Website Security

Website security is more important than ever. Web servers, which host the data and other content available to your customers on the Internet, are often the most targeted and attacked components of a company's network. Cyber criminals are constantly looking for improperly secured websites to attack, while many customers say website security is a top consideration when they choose to shop online. As a result, it is essential to secure servers and the network infrastructure that supports them. The consequences of a security breach are great: loss of revenues, damage to credibility, legal liability and loss of customer trust.

The following are examples of specific security threats to web servers:

- Cyber criminals may exploit software bugs in the web server, underlying operating system, or active content to gain unauthorized access to the web server. Examples of unauthorized access include gaining access to files or folders that were not meant to be publicly accessible and being able to execute commands and/or install malicious software on the web server.
- Denial-of-service attacks may be directed at the web server or its supporting network infrastructure to prevent or hinder your website users from making use of its services. This can include preventing the user from accessing email, websites, online accounts or other services. The most common attack occurs when the attacker floods a network with information, so that it can't process the user's request.
- Sensitive information on the web server may be read or modified without authorization.
- Sensitive information on backend databases that are used to support interactive elements of a web application may be compromised through the injection of unauthorized software commands. Examples include Structured Query Language (SQL) injection, Lightweight Directory Access Protocol (LDAP) injection and cross-site scripting (XSS).
- Sensitive unencrypted information transmitted between the web server and the browser may be intercepted.
- Information on the web server may be changed for malicious purposes. Website defacement is a commonly reported example of this threat.
- Cyber criminals may gain unauthorized access to resources elsewhere in the organization's network via a successful attack on the web server.
- Cyber criminals may also attack external entities after compromising a web server. These attacks can be launched directly (e.g., from the compromised server against an external server) or indirectly (e.g., placing malicious content on the compromised web server that attempts to exploit vulnerabilities in the web browsers of users visiting the site).
- The server may be used as a distribution point for attack tools, pornography or illegally copied software.

Cyber Plan Action Items:

1. Carefully plan and address the security aspects of the deployment of a public web server.

Because it is much more difficult to address security once deployment and implementation have occurred, security should be considered from the initial planning stage. Businesses are more likely to make decisions about configuring computers appropriately and consistently when they develop and use a detailed, well-designed deployment plan. Developing such a plan will support web server administrators in making the inevitable tradeoff decisions between usability, performance and risk.

Businesses also need to consider the human resource requirements for the deployment and continued operation of the web server and supporting infrastructure. The following points in a deployment plan:

- Types of personnel required -- for example, system and web server administrators, webmasters, network administrators and information systems security personnel.

- Skills and training required by assigned personnel.
- Individual (i.e., the level of effort required of specific personnel types) and collective staffing (i.e., overall level of effort) requirements.

2. Implement appropriate security management practices and controls when maintaining and operating a secure web server.

Appropriate management practices are essential to operating and maintaining a secure web server. Security practices include the identification of your company's information system assets and the development, documentation and implementation of policies, and guidelines to help ensure the confidentiality, integrity and availability of information system resources. The following practices and controls are recommended:

- A business-wide information system security policy.
- Server configuration and change control and management.
- Risk assessment and management.
- Standardized software configurations that satisfy the information system security policy.
- Security awareness and training.
- Contingency planning, continuity of operations and disaster recovery planning.
- Certification and accreditation.

3. Ensure that web server operating systems meet your organization's security requirements.

The first step in securing a web server is securing the underlying operating system. Most commonly available web servers operate on a general-purpose operating system. Many security issues can be avoided if the operating systems underlying web servers are configured appropriately. Default hardware and software configurations are typically set by manufacturers to emphasize features, functions and ease of use at the expense of security. Because manufacturers are not aware of each organization's security needs, each web server administrator must configure new servers to reflect their business' security requirements and reconfigure them as those requirements change. Using security configuration guides or checklists can assist administrators in securing systems consistently and efficiently. Initially securing an operating system initially generally includes the following steps:

- Patch and upgrade the operating system.
- Change all default passwords
- Remove or disable unnecessary services and applications.
- Configure operating system user authentication.
- Configure resource controls.
- Install and configure additional security controls.
- Perform security testing of the operating system.

4. Ensure the web server application meets your organization's security requirements.

In many respects, the secure installation and configuration of the web server application will mirror the operating system process discussed above. The overarching principle is to install the minimal amount of web server services required and eliminate any known vulnerabilities through patches or upgrades. If the installation program installs any unnecessary applications, services or scripts, they should be removed immediately after the installation process concludes. Securing the web server application generally includes the following steps:

- Patch and upgrade the web server application.
- Remove or disable unnecessary services, applications and sample content.

- Configure web server user authentication and access controls.
- Configure web server resource controls.
- Test the security of the web server application and web content.

5. Ensure that only appropriate content is published on your website.

Company websites are often one of the first places cyber criminals search for valuable information. Still, many businesses lack a web publishing process or policy that determines what type of information to publish openly, what information to publish with restricted access and what information should not be published to any publicly accessible repository. Some generally accepted examples of what should not be published or at least should be carefully examined and reviewed before being published on a public website include:

- Classified or proprietary business information.
- Sensitive information relating to your business' security.
- Medical records.
- A business' detailed physical and information security safeguards.
- Details about a business' network and information system infrastructure -- for example, address ranges, naming conventions and access numbers.
- Information that specifies or implies physical security vulnerabilities.
- Detailed plans, maps, diagrams, aerial photographs and architectural drawings of business buildings, properties or installations.
- Any sensitive information about individuals that might be subject to federal, state or, in some instances, international privacy laws.

6. Ensure appropriate steps are taken to protect web content from unauthorized access or modification.

Although information available on public websites is intended to be public (assuming a credible review process and policy is in place), it is still important to ensure that information cannot be modified without authorization. Users of such information rely on its integrity even if the information is not confidential. Content on publicly accessible web servers is inherently more vulnerable than information that is inaccessible from the Internet, and this vulnerability means businesses need to protect public web content through the appropriate configuration of web server resource controls. Examples of resource control practices include:

- Install or enable only necessary services.
- Install web content on a dedicated hard drive or logical partition.
- Limit uploads to directories that are not readable by the web server.
- Define a single directory for all external scripts or programs executed as part of web content.
- Disable the use of hard or symbolic links.
- Define a complete web content access matrix identifying which folders and files in the web server document directory are restricted, which are accessible, and by whom.
- Disable directory listings.
- Deploy user authentication to identify approved users, digital signatures and other cryptographic mechanisms as appropriate.
- Use intrusion detection systems, intrusion prevention systems and file integrity checkers to spot intrusions and verify web content.
- Protect each backend server (i.e., database server or directory server) from command injection attacks.

7. Use active content judiciously after balancing the benefits and risks.

Static information resided on the servers of most early websites, typically in the form of text-based documents. Soon thereafter, interactive elements were introduced to offer new opportunities for user interaction.

Unfortunately, these same interactive elements introduced new web-related vulnerabilities. They typically involve dynamically executing code using a large number of inputs, from web page URL parameters to hypertext transfer protocol (HTTP) content and, more recently, extensible markup language (XML) content. Different active content technologies pose different related vulnerabilities, and their risks should be weighed against their benefits. Although most websites use some form of active content generators, many also deliver some or all of their content in a static form.

8. Use authentication and cryptographic technologies as appropriate to protect certain types of sensitive data.

Public web servers often support technologies for identifying and authenticating users with differing privileges for accessing information. Some of these technologies are based on cryptographic functions that can provide a secure channel between a web browser client and a web server that supports encryption. Web servers may be configured to use different cryptographic algorithms, providing varying levels of security and performance.

Without proper user authentication in place, businesses cannot selectively restrict access to specific information. All information that resides on a public web server is then accessible by anyone with access to the server. In addition, without some process to authenticate the server, users of the public web server will not be able to determine whether the server is the "authentic" web server or a counterfeit version operated by a cyber criminal.

Even with an encrypted channel and an authentication mechanism, it is possible that attackers may attempt to access the site by brute force. Improper authentication techniques can allow attackers to gather valid usernames or potentially gain access to the website. Strong authentication mechanisms can also protect against phishing attacks, in which hackers may trick users into providing their personal credentials, and pharming, in which traffic to a legitimate website may be redirected to an illegitimate one. An appropriate level of authentication should be implemented based on the sensitivity of the web server's users and content.

9. Employ network infrastructure to help protect public web servers.

The network infrastructure (e.g., firewalls, routers, intrusion detection systems) that supports the web server plays a critical security role. In most configurations, the network infrastructure will be the first line of defense between a public web server and the Internet. Network design alone, though, cannot protect a web server. The frequency, sophistication and variety of web server attacks perpetrated today support the idea that web server security must be implemented through layered and diverse protection mechanisms, an approach sometimes referred to as "defense-in-depth."

10. Commit to an ongoing process of maintaining web server security.

Maintaining a secure web server requires constant effort, resources and vigilance. Securely administering a web server on a daily basis is essential. Maintaining the security of a web server will usually involve the following steps:

- Configuring, protecting and analyzing log files.
- Backing up critical information frequently.
- Maintaining a protected authoritative copy of your organization's web content.
- Establishing and following procedures for recovering from compromise.
- Testing and applying patches in a timely manner.

- Testing security periodically.

Email

Email has become a critical part of our everyday business, from internal management to direct customer support. The benefits associated with email as a primary business tool far outweigh the negatives. However, businesses must be mindful that a successful email platform starts with basic principles of email security to ensure the privacy and protection of customer and business information.

Cyber Plan Action Items:

1. Set up a spam email filter

It has been well documented that spam, phishing attempts and otherwise unsolicited and unwelcome email often accounts for more than 60 percent of all email that an individual or business receives. Email is the primary method for spreading viruses and malware and it is one of the easiest to defend against. Consider using email-filtering services that your email service, hosting provider or other cloud providers offer. A local email filter application is also an important component of a solid antivirus strategy. Ensure that automatic updates are enabled on your email application, email filter and anti-virus programs. Ensure that filters are reviewed regularly so that important email and/or domains are not blocked in error.

2. Train your employees in responsible email usage

The last line of defense for all of your cyber risk efforts lies with the employees who use tools such as email and their responsible and appropriate use and management of the information under their control. Technology alone cannot make a business secure. Employees must be trained to identify risks associated with email use, how and when to use email appropriate to their work, and when to seek assistance of professionals. Employee awareness training is available in many forms, including printed media, videos and online training.

Consider requiring security awareness training for all new employees and refresher courses every year. Simple efforts such as monthly newsletters, urgent bulletins when new viruses are detected, and even posters in common areas to remind your employees of key security and privacy to-do's create a work environment that is educated in protecting your business.

3. Protect sensitive information sent via email

With its proliferation as a primary tool to communicate internally and externally, business email often includes sensitive information. Whether it is company information that could harm your business or regulated data such as personal health information (PHI) or personally identifiable information (PII), it is important to ensure that such information is only sent and accessed by those who are entitled to see it.

Since email in its native form is not designed to be secure, incidents of misaddressing or other common accidental forwarding can lead to data leakage. Businesses that handle this type of information should consider whether such information should be sent via email, or at least consider using email encryption. Encryption is the process of converting data into unreadable format to prevent disclosure to unauthorized personnel. Only individuals or organizations with access to the encryption key can read the information. Other cloud services offer "Secure Web Enabled Drop Boxes" that enable secure data transfer for sensitive information, which is often a better approach to transmitting between companies or customers.

4. Set a sensible email retention policy

Another important consideration is the management of email that resides on company messaging systems and your users' computers. From the cost of storage and backup to legal and regulatory requirements, companies should document how they will handle email retention and implement basic controls to help them attain those standards. Many industries have specific rules that dictate how long emails can or should be retained, but the basic rule of thumb is only as long as it supports your business efforts. Many companies implement a 60-90 day retention standard if not compelled by law to another retention period.

To ensure compliance, companies should consider mandatory archiving at a chosen retention cycle end date and automatic permanent email removal after another set point, such as180-360 days in archives. In addition, organizations should discourage the use of personal folders on employee computers (most often configurable from the e-mail system level), as this will make it more difficult to manage company standards.

5. Develop an email usage policy

Policies are important for setting expectations with your employees or users, and for developing standards to ensure adherence to your published polices.

Your policies should be easy to read, understand, define and enforce. Key areas to address include what the company email system should and should not be used for, and what data are allowed to be transmitted. Other policy areas should address retention, privacy and acceptable use.

Depending on your business and jurisdiction, you may have a need for email monitoring. The rights of the business and the user should be documented in the policy as well. The policy should be part of your general end user-awareness training and reviewed for updates on a yearly basis.

For a sample email usage policy, see: http://www.sans.org/security-resources/policies/Email_Policy.pdf

Mobile Devices

If your company uses mobile devices to conduct company business, such as accessing company email or sensitive data, pay close attention to mobile security and the potential threats that can expose and compromise your overall business networks. This section describes the mobile threat environment and the practices that small businesses can use to help secure devices such as smartphones, tablets and Wi-Fi enabled laptops.

Many organizations are finding that employees are most productive when using mobile devices, and the benefits are too great to ignore. But while mobility can increase workplace productivity, allowing employees to bring their own mobile devices into the enterprise can create significant security and management challenges.

Data loss and data breaches caused by lost or stolen phones create big challenges, as mobile devices are now used to store confidential business information and access the corporate network. According to a December 2010 Symantec mobile security survey, 68 percent of respondents ranked loss or theft as their top mobile-device security concern, while 56 percent said mobile malware is their number two concern. It is important to remember that while the individual employee may be liable for a device, the company is still liable for the data.

Top threats targeting mobile devices

- *Data Loss* – An employee or hacker accesses sensitive information from device or network. This can be unintentional or malicious, and is considered the biggest threat to mobile devices

- *Social Engineering Attacks* – A cyber criminal attempts to trick users to disclose sensitive information or install malware. Methods include phishing and targeted attacks.

- *Malware* – Malicious software that includes traditional computer viruses, computer worms and Trojan horse programs. Specific examples include the Ikee worm, targeting iOS-based devices; and Pjapps malware that can enroll infected Android devices in a collection of hacker-controlled "zombie" devices known as a "botnet."

- *Data Integrity Threats* – Attempts to corrupt or modify data in order to disrupt operations of a business for financial gain. These can also occur unintentionally.

- *Resource Abuse* – Attempts to misuse network, device or identity resources. Examples include sending spam from compromised devices or denial of service attacks using computing resources of compromised devices.

- *Web and Network-based Attacks* – Launched by malicious websites or compromised legitimate sites, these target a device's browser and attempt to install malware or steal confidential data that flows through it.

Cyber Plan Action Items:

A few simple steps can to help ensure company information is protected. These include requiring all mobile devices that connect to the business network be equipped with security software and password protection; and providing general security training to make employees aware of the importance of security practices for mobile devices. More specific practices are detailed below.

1. Use security software on all smartphones

Security software specifically designed for smartphones can stop hackers and prevent cyber criminals from stealing your information or spying on you when you use public networks. It can detect and remove viruses and other mobile threats before they cause you problems. It can also eliminate annoying text and multimedia spam messages.

2. Make sure all software is up to date

Mobile devices must be treated like personal computers in that all software on the devices should be kept current, especially the security software. This will protect devices from new variants of malware and viruses that threaten your company's critical information.

3. Encrypt the data on mobile devices

Business and personal information stored on mobile devices is often sensitive. Encrypting this data is another must. If a device is lost and the SIM card stolen, the thief will not be able to access the data if the proper encryption technology is loaded on the device.

4. Have users password protect access to mobile devices

In addition to encryption and security updates, it is important to use strong passwords to protect data stored on mobile devices. This will go a long way toward keeping a thief from accessing sensitive data if the device is lost or hacked.

5. Urge users to be aware of their surroundings

Whether entering passwords or viewing sensitive or confidential data, users should be cautious of who might be looking over their shoulder.

6. Employ these strategies for email, texting and social networking

Avoid opening unexpected text messages from unknown senders – As with email, attackers can use text messages to spread malware, phishing scams and other threats among mobile device users. The same caution should be applied to opening unsolicited text messages that users have become accustomed to with email.

Don't be lured in by spammers and phishers – To shield business networks from cyber criminals, small businesses should deploy appropriate email security solutions, including spam prevention, which protect a company's reputation and manage risks.

Click with caution – Just like on stationary PCs, social networking on mobile devices and laptops should be conducted with care and caution. Users should not open unidentified links, chat with unknown people or visit unfamiliar sites. It doesn't take much for a user to be tricked into compromising a device and the information on it.

7. Set reporting procedures for lost or stolen equipment

In the case of a loss or theft, employees and management should all know what to do next. Processes to deactivate the device and protect its information from intrusion should be in place. Products are also available for the automation of such processes, allowing small businesses to breathe easier after such incidents.

8. Ensure all devices are wiped clean prior to disposal

Most mobile devices have a reset function that allows all data to be wiped. SIM cards should also be removed and destroyed.

Helpful links:

- Teach your employees about mobile apps:
  http://onguardonline.gov/articles/0018-understanding-mobile-apps
- Keep your laptops secure:
  http://onguardonline.gov/articles/0015-laptop-security

Employees

Businesses must establish formal recruitment and employment processes to control and preserve the quality of their employees. Many employers have learned the hard way that hiring someone with a criminal record, falsified credentials or undesirable background can create a legal and financial nightmare.

Without exercising due diligence in hiring, employers run the risk of making unwise hiring choices that can lead to workplace violence, theft, embezzlement, lawsuits for negligent hiring and numerous other workplace problems.

Cyber Plan Action Items:

1. Develop a hiring process that properly vets candidates

The hiring process should be a collaborative effort among different groups of your organization, including recruitment, human resources, security, legal and management teams. It is important to have a solid application, resume, interview and reference-checking process to identify potential gaps and issues that may appear in a background check.

An online employment screening resource called the "Online Safe Hiring Certification Course" can help you set the groundwork for a safe recruitment process. The course will teach your teams what to look for in the different stages of the hiring process, how to interview and how to set up a safe hiring program to avoid hiring an employee that may be problematic. The course is available here: http://www.esrcheck.com/ESRonlineSafeHiringCourse.php.

2. Perform background checks and credentialing

Background checks are essential and must be consistent. Using a background screening company is highly recommended. The standard background screening should include the following checks:

- Employment verification
- Education verification
- Criminal records
- Drug testing
- The U.S. Treasury Office of Foreign Affairs and Control
- Sex offender registries
- Social Security traces and validation

Depending on the type of your business, other screening criteria may consist of credit check, civil checks and federal criminal checks. Conducting post-hire checks for all employees every two to three years, depending on your industry, is also recommended.

If you do conduct background checks, you as an employer have obligations under the Fair Credit Reporting Act. For more information about employer obligations under the FCRA, visit http://business.ftc.gov/documents/bus08-using-consumer-reports-what-employers-need-know.

3. Take care in dealing with third parties

Employers should properly vet partner companies through which your organization hires third-party consultants. To ensure consistent screening criteria are enforced for third-party consultants, you need to explicitly set the credentialing requirements in your service agreement. State in the agreement that the company's credentialing requirements must be followed.

4. Set appropriate access controls for employees

Both client data and internal company data are considered confidential and need particular care when viewed, stored, used, transmitted or disposed. It is important to analyze the role of each employee and set data access control based upon the role. If a role does not require the employee to ever use sensitive data, the employee's access to the data should be strictly prohibited. However, if the role requires the employee to work with sensitive data, the level of access must be analyzed thoroughly and be assigned in a controlled and tiered manner following "least-privilege" principles, which allow the employee to only access data that is necessary to perform his or her job.

If the organization does not have a system in place to control data access, the following precautions are strongly recommended. Every employee should:

- Never access or view client data without a valid business reason. Access should be on a need-to-know basis.
- Never provide confidential data to anyone – client representatives, business partners or even other employees – unless you are sure of the identity and authority of that person.
- Never use client data for development, testing, training presentations or any purpose other than providing production service, client-specific testing or production diagnostics. Only properly sanitized data that cannot be traced to a client, client employee, customer or your organization's employee should be used for such purposes.
- Always use secure transmission methods such as secure email, secure file transfer (from application to application) and encrypted electronic media (e.g., CDs, USB drives or tapes).
- Always keep confidential data (hard copy and electronic) only as long as it is needed.
- Follow a "clean desk" policy, keeping workspaces uncluttered and securing sensitive documents so that confidential information does not get into the wrong hands.
- Always use only approved document disposal services or shred all hardcopy documents containing confidential information when finished using them. Similarly, use only approved methods that fully remove all data when disposing of, sending out for repair or preparing to reuse electronic media.

5. Provide security training for employees

Security awareness training teaches employees to understand system vulnerabilities and threats to business operations that are present when using a computer on a business network.

A strong IT security program must include training IT users on security policy, procedures and techniques, as well as the various management, operational and technical controls necessary and available to keep IT resources secure. In addition, IT infrastructure managers must have the skills necessary to carry out their assigned duties effectively. Failure to give attention to the area of security training puts an enterprise at great risk because security of business resources is as much a human issue as it is a technology issue.

Technology users are the largest audience in any organization and are the single most important group of people who can help to reduce unintentional errors and IT vulnerabilities. Users may include employees, contractors, foreign or domestic guest researchers, other personnel, visitors, guests and other collaborators or associates requiring access. Users must:

- Understand and comply with security policies and procedures.
- Be appropriately trained in the rules of behavior for the systems and applications to which they have access.
- Work with management to meet training needs.
- Keep software and applications updated with security patches.
- Be aware of actions they can take to better protect company information. These actions include: proper password usage, data backup, proper antivirus protection, reporting any suspected incidents or violations of

security policy, and following rules established to avoid social engineering attacks and deter the spread of spam or viruses and worms.

A clear categorization of what is considered sensitive data versus non-sensitive data is also needed. Typically, the following data are considered sensitive information that should be handled with precaution:

- Government issued identification numbers (e.g., Social Security numbers, driver's license numbers)
- Financial account information (bank account numbers, credit card numbers)
- Medical records
- Health insurance information
- Salary information
- Passwords

The training should cover security policies for all means of access and transmission methods, including secure databases, email, file transfer, encrypted electronic media and hard copies.

Employers should constantly emphasize the critical nature of data security. Regularly scheduled refresher training courses should be established in order to instill the data security culture of your organization. Additionally, distribute data privacy and security related news articles in your training, and send organization-wide communication on notable data privacy related news as reminders to your employees.

6. Implement Employee Departure Checklist

Create a security checkout checklist for employees that are no longer with your company, regardless of their reason for leaving (voluntary or involuntary). It's recommended by the U.S. Chamber of Commerce and others that all small businesses ensure terminated employee accounts are erased on all network devices and drives immediately. This is especially true for any devices that may have been taken offsite such as laptops and smartphones.

Helpful links

- Stop.Think.Connect. Internal Employee Rollout Materials
  http://www.dhs.gov/stopthinkconnect
- Internet Safety at Work PowerPoint Presentation
  http://go microsoft.com/?linkid=9745638
- Tip Cards: Top Tips for Internet Safety at Work
- http://go microsoft.com/?linkid=9745642
- Video: "Stay Sharp on Internet Safety at Work"
  http://go microsoft.com/?linkid=9745640
- U.S. Chamber of Commerce: Internet Security Essentials for Business 2.0
  https://www.uschamber.com/sites/default/files/issues/technology/files/ISEB-2.0-CyberSecurityGuide.pdf

Facility Security

Protecting employees and members of the public who visit your facility is a complex and challenging responsibility. It's also one of your company's top priorities.

Cyber Plan Action Items:

1. Recognize the importance of securing your company facilities

The physical security of a facility depends on a number of security decisions that can be identified through a comprehensive risk-management process. The objective of risk management is to identify an achievable level of protection for your company that corresponds as closely as possible to the level of risk without exceeding the risk.

It is easy to think about physical security of your company's facility as merely an exercise in maintaining control of access points and ensuring there is complete visibility in areas that are determined to be of high-risk – either because of the threat of easy public access or because of the value of information located nearby. However, maintaining security of your company's facility also includes the physical environment of public spaces. For instance:

- Employees whose computers have access to sensitive information should not have their computer monitors oriented toward publicly accessible spaces such as reception areas, check-in desks and waiting rooms. Employees should be trained to not write out logins and passwords on small pieces of paper affixed to computer equipment viewable in public spaces.
- Easy-to-grab equipment that could contain sensitive or personally identifiable information – such as laptops, electronic tablets and cell phones – should be located away from public areas. If you have an environment where employees are working in a waiting room or reception area, train them to not leave these types of devices out on their desks unsecured.
- Consider using cable locks as an easy way to increase security for laptop computers. Most laptops feature a lock port for a cable which can be connected to the user's desk. Be sure to store the key to the cable lock in a secure location away from the desk the computer is locked to.
- In cases that extremely sensitive information is stored on a laptop, consider adding a LoJack software system. The software runs unnoticed and allows law enforcement to locate stolen computers more easily and also allows an administrator to wipe the hard drive remotely if necessary.
- Consider implementing a badge identification system for all employees, and train employees to stop and question anyone in the operational business area without a badge or who appears to be an unescorted visitor.

2. Minimize and safeguard printed materials with sensitive information

Probably the most effective way to minimize the risk of losing control of sensitive information from printed materials is to minimize the amount of printed materials that contain sensitive information. Management procedures should limit how many instances and copies of printed reports memoranda and other material containing personally identifiable information exist.

Safeguard copies of material containing sensitive information by providing employees with locking file cabinets or safes. Make it a standard operating procedure to lock up important information. Train employees to understand that simply leaving the wrong printed material on a desk, in view of the general public, can result in consequences that impact the entire company and your customers.

3. Ensure mail security

Your mail center can introduce a wide range of potential threats to your business. Your center's screening and handling processes must be able to identify threats and hoaxes and eliminate or mitigate the risk they pose to facilities, employees and daily operations. Your company should ensure that mail managers understand the range of screening procedures and evaluate them in terms of your specific operational requirements.

4. Dispose of trash securely

Too often, sensitive information – including customers' personally identifiable information, business financial and other data, and company system access information – is available for anyone to find in the trash. Invest in business-grade shredders and buy enough of them to make it convenient for employees. Alternatively, subscribe to a trusted shredding company that will provide locked containers for storage until documents are shredded. Develop standard procedures and employee training programs to ensure that everyone in your company is aware of what types of information need to be shredded.

5. Dispose electronic equipment securely

Be aware that emptying the recycle bin on your desktop or deleting documents from folders on your computer or other electronic device may not delete information forever. Those with advanced computer skills can still access your information even after you think you've destroyed it.

Disposing of electronic equipment requires skilled specialists in order to ensure the security of sensitive information contained within that equipment. If outside help, such as an experienced electronic equipment recycler and data security vendor, is not available or too expensive, you should at a minimum remove computer hard drives and have them shredded. Also, be mindful of risks with other types of equipment associated with computer equipment, including CDs and thumb drives.

6. Train your employees in facility security procedures

A security breach of customer information or a breach of internal company information can result in a public loss of confidence in your company and can be as devastating for your business as a natural disaster. In order to address such risks, you must devote your time, attention and resources (including employee training time) to the potential vulnerabilities in your business environment and the procedures and practices that must be a standard part of each employee's workday.

And while formal training is important to maintaining security, the daily procedures you establish in both the normal conduct of business and in the way you model good security behaviors and practices are equally important. In short, security training should be stressed as critical and reinforced via daily procedures and leadership modeling.

Operational Security

While operational security, or OPSEC, has its origins in securing information important to military operations, it has applications across the business community today.

In a commercial context, OPSEC is the process of denying hackers access to any information about the capabilities or intentions of a business by identifying, controlling and protecting evidence of the planning and execution of activities that are essential the success of operations.

OPSEC is a continuous process that consists of five distinct actions:

- Identify information that is critical to your business.
- Analyze the threat to that critical information.
- Analyze the vulnerabilities to your business that would allow a cyber criminal to access critical information.
- Assess the risk to your business if the vulnerabilities are exploited.
- Apply countermeasures to mitigate the risk factors.

In addition to being a five-step process, OPSEC is also a mindset that all business employees should embrace. By educating oneself on OPSEC risks and methodologies, protecting sensitive information that is critical to the success of your business becomes second nature.

This section explains the OPSEC process and provides some general guidelines that are applicable to most businesses. An understanding of the following terms is required before the process can be explained:

- *Critical information* – Specific data about your business strategies and operations that are needed by cyber criminals to hamper or harm your business from successfully operating.
- *OPSEC indicators* – Business operations and publicly available information that can be interpreted or pieced together by a cyber criminal to derive critical information.
- *OPSEC vulnerability* – A condition in which business operations provide OPSEC indicators that may be obtained and accurately evaluated by a cyber criminal to provide a basis for hampering or harming successful business operations.

Cyber Plan Action Items:

1. Identity of critical information

The identification of critical information is important in that it focuses the remainder of the OPSEC process on protecting vital information rather than attempting to protect all information relevant to business operations. Given that any business has limited time, personnel and money for developing secure business practices, it is essential to focus those limited resources on protecting information that is most critical to successful business operations. Examples of critical information include, but should not be limited to, the following:

- Customer lists and contact information
- Contracts
- Patents and intellectual property
- Leases and deeds
- Policy manuals
- Articles of incorporation
- Corporate papers
- Laboratory notebooks

- Audio tapes
- Video tapes
- Photographs and slides
- Strategic plans and board meeting minutes

Importantly, what is critical information for one business may not be critical for another business. Use your company's mission as a guide for determining what data are truly vital.

2. Analyze threats

This action involves research and analysis to identify likely cyber criminals who may attempt to obtain critical information regarding your company's operations. OPSEC planners in your business should answer the following critical information questions:

- Who might be a cyber criminal (e.g. competitors, politically motivated hackers, etc.)?
- What are the cyber criminal's goals?
- What actions might the cyber criminal take?
- What critical information does the cyber criminal already have on your company's operations? (i.e., what is already publicly available?)

3. Analyze vulnerabilities

The purpose of this action is to identify the vulnerabilities of your business in protecting critical information. It requires examining each aspect of security that seeks to protect your critical information and then comparing those indicators with the threats identified in the previous step. Common vulnerabilities for small businesses include the following:

- Poorly secured mobile devices that have access to critical information.
- Lack of policy on what information and networked equipment can be taken home from work or taken abroad on travel.
- Storage of critical information on personal email accounts or other non-company networks.
- Lack of policy on what business information can be posted to or accessed by social network sites.

4. Assess risk

This action has two components. First, OPSEC managers must analyze the vulnerabilities identified in the previous action and identify possible OPSEC measures to mitigate each one. Second, specific OPSEC measures must be selected for execution based upon a risk assessment done by your company's senior leadership. Risk assessment requires comparing the estimated cost associated with implementing each possible OPSEC measure to the potential harmful effects on business operations resulting from the exploitation of a particular vulnerability.

OPSEC measures may entail some cost in time, resources, personnel or interference with normal operations. If the cost to achieve OPSEC protection exceeds the cost of the harm that an intruder could inflict, then the application of the measure is inappropriate. Because the decision not to implement a particular OPSEC measure entails risks, this step requires your company's leadership approval.

5. Apply appropriate OPSEC measures

In this action, your company's leadership reviews and implements the OPSEC measures selected in the assessment of risk action. Before OPSEC measures can be selected, security objectives and critical information must be known, indicators identified and vulnerabilities assessed.

Helpful links

These resources provide additional information on the origins, purpose and implementation of operational security.

- National Security Agency/Central Security Service, *PURPLE DRAGON: The Origin and Development of the United States OPSEC Program* (1993):
  http://www.nsa.gov/public_info/_files/cryptologic_quarterly/purple_dragon.pdf
- Joint Publication 3-13.3, *Operations Security* (29 June 2006): Available through Joint Doctrine Education and Training Electronic Information System (JDEIS). http://www.iad.gov/ioss/media/documents/Joint_Pub_3-13-3.pdf
- National OPSEC Program: https://www.iad.gov/ioss/
- OPSEC Professionals Society: http://opsecsociety.org/
- Operations Security Professional's Association: http://www.opsecprofessionals.org/
- Department of Homeland Security Critical Infrastructure Protection: http://www.dhs.gov/criticalinfrastructure

Payment Cards

If your business accepts payment by credit or debit cards, it is important to have security steps in place to ensure your customer information is safe. You also may have security obligations pursuant to agreements with your bank or payment services processor. These entities can help you prevent fraud. In addition, free resources and general security tips are available to learn how to keep sensitive information – beyond payment information – safe.

Cyber Plan Action Items:

1. Understand and catalog customer and card data you keep

- Make a list of the type of customer and card information you collect and keep – names, addresses, identification information, payment card numbers, magnetic stripe data, bank account details and Social Security numbers. It's not only card numbers criminals want; they're looking for all types of personal information, especially if it helps them commit identity fraud.
- Understand where you keep such information and how it is protected.
- Determine who has access to this data and if they need to have access.

2. Evaluate whether you need to keep all the data you store

- Once you know what information you collect and store, evaluate whether you really need to keep it. Often businesses may not realize they're logging or otherwise keeping unnecessary data until they conduct an audit. Not keeping sensitive data in storage makes it harder for criminals to steal it.
- If you've been using card numbers for purposes other than payment transactions, such as a customer loyalty program, ask your merchant processor if you can use alternative data instead. Tokenization, for example, is technology that masks card numbers and replaces it with an alternate number that can't be used for fraud.

3. Use secure tools and services

- The payments industry maintains lists of hardware, software and service providers who have been validated against industry security requirements.
- Small businesses that use integrated payment systems, in which the card terminal is connected to a larger computer system, can check the list of validated payment applications to make sure any software they employ has been tested.
- Have a conversation about security with your provider if the products or services you are currently using are not on the lists.

4. Control access to payment systems

- Whether you use a more complicated payment system or a simple standalone terminal, make sure you carefully control access.
- Isolate payment systems from other, less secure programs, especially those connected to the Internet. For example, don't use the same computer to process payments and surf the Internet.
- Control or limit access to payment systems to only employees who need access.
- Make sure you use a secure system for remote access or eliminate remote access if you don't need it so that criminals cannot infiltrate your system from the Internet.

5. Use security tools and resources

Work with your bank or processor and ask about the anti-fraud measures, tools and services you can use to ensure criminals cannot use stolen card information at your business.

- For e-commerce retailers:
  - The CVV2 code is the three-digit number on the signature panel that can help verify that the customer has physical possession of the card and not just the account number.
  - Retailers can also use Address Verification Service to ensure the cardholder has provided the correct billing address associated with the account.
  - Services such as Verified by Visa prompt the cardholder to enter a personal password confirming their identity and providing an extra layer of protection.
- For brick and mortar retailers:
  - Swipe the card and get an electronic authorization for the transaction.
  - Check that the signature matches the card.
  - Ensure your payment terminal is secure and safe from tampering.

6. Remember the security basics

- Use strong, unique passwords and change them frequently.
- Use up-to-date firewall and anti-virus technologies.
- Do not click on suspicious links you may receive by email or encounter online.

Helpful links

You don't have to tackle security on your own. Work with your bank or processor to make sure you're getting the support and expertise you need.

- Visa offers a data security guide for small business as part of its Cardholder Information Security Program: http://usa.visa.com/merchants/risk management/data security demo/popup.html
- Information about industry security standards is available from the PCI Security Standards Council: https://www.pcisecuritystandards.org
- The Paysimple.com blog offers a helpful post on credit card security: http://paysimple.com/blog/2011/09/01/5-tips-for-proper-handling-of-customer-credit-card-account-information/
- American Express provides data security advice for merchants: https://www260.americanexpress.com/merchant/singlevoice/dsw/FrontServlet?request type=dsw&pg nm=merchinfo&ln=en&frm=US
- MasterCard offers resources for on safeguarding customer information. : http://www mastercard.com/us/business/en/smallbiz/resources/industry/e-commerce/articles/0802CustomerData html

Incident Response

Even well-implemented cyber security structures and plans may not prevent all breaches of your business' data defenses, so be sure to have procedures in place to respond to security breaches when they occur.

Types of breaches

Physical breaches include real-world crimes such as burglaries and equipment theft, as well as any event when your company's equipment is misplaced or lost in transit. Unauthorized devices may be installed on a system or network, permitting further compromises of data confidentiality and integrity. Physical breaches can also result from reselling, donating or recycling old equipment that has not been properly cleansed of potentially sensitive information.

Network and system security breaches include events when computers become infected with malicious code, are accessed by unauthorized individuals remotely or are used by authorized individuals to perform malicious activity. This can also include breaches to network routers and firewalls, both within and outside your organization's boundary and control.

Data breaches, meaning the leakage or spillage of sensitive information into insecure channels, can result from any of the types of events described above. Data breaches can also occur if sensitive information is left improperly exposed by mistake.

Cyber Plan Action Items, if Breach Occurs:

1. Notify law enforcement if necessary

Depending on the type of breach and type of business, your company may be required to notify local law enforcement or other government authorities upon discovery of a data breach. In the event of exposure of customer information, you should notify the customer(s) of the incident, record the data that was lost or exposed and record the measures taken to ensure against future exposure.

2. Work cohesively across technical and leadership teams to limit the damage

Once your company becomes aware that a breach has occurred, technical personnel and business decision makers should work together to decide on the most practical and effective containment plan. Containment plans will vary from one set of circumstances to the next, and they may quickly become intensive in terms of time and resources from both the technological and business impact perspectives. In any case, the containment of data breaches should be focused on determining the extent of the compromise and preserving the confidentiality and integrity of sensitive data that has not yet been stolen or disclosed.

Other issues affecting the selection and execution of a containment plan include your company's reputation-risk management strategy and the decision on whether to request outside assistance – either from local or federal law enforcement, a private consulting firm or a computer incident response organization such as US-CERT.

3. Begin recovery effort

After a containment plan has been established and execution has begun, get started on eradication and recovery efforts. In the case of network and system security breaches, eradication usually means removing all instances of unauthorized software from the network and disabling all access privileges associated with users who have committed malicious activity.

Cleaning a network or system of all traces of malicious code can often entail having to completely wipe all storage media and perform a "clean install." Therefore, recovery from such a breach may be resource intensive and require careful restoration of data from backups. Bear in mind that backups may also contain malicious code and should be carefully checked for compromise; otherwise, the security breach will be perpetuated after the recovery phase.

Key Disaster Recovery Principles

- *Don't wait until it's too late* – Small businesses should not wait until after a disaster to think about what should have been done to protect their data. Not only is downtime costly from a financial perspective, but it could mean the complete demise of the business. Small businesses should map out disaster preparedness plans ahead of time, including the identification of key systems, data and other resources that are critical to running the business.

- *Protect information completely* – To reduce the risk of losing critical business information, small businesses must implement the appropriate security and backup solutions to archive important files, such as customer records and financial information for the long term. Natural disasters, theft and cyber attacks can all result in data and financial loss, so small businesses need to make sure important files are saved not only on an external hard drive and/or company network, but in a safe, off-site location.

- *Get employees involved* – Employees play a key role in helping to prevent downtime. They should be educated on computer security best practices and what to do if information is accidentally deleted or cannot easily be found in their files. Since small businesses often have limited resources, all employees should know how to retrieve the businesses' information in times of disaster.

- *Test frequently* – After a disaster hits is the worst time to learn that critical files were not backed up as planned. Regular disaster recovery testing is invaluable. Test your plan anytime anything changes in your environment.

- *Review your plan* – If frequent testing is not feasible due to resources and bandwidth, small businesses should at least review disaster preparedness plan on a quarterly basis.

- *Be prepared* – It is always better and less costly to invest in adequate security up-front rather than going through a costly incident response which could result in rebuilding your entire network infrastructure.

4. Hold a 'lessons learned' meeting

Lastly, your company should always perform a "lessons learned" meeting after the recovery phase has been successfully completed to discover, document and refine the knowledge gained during the incident handling process.

Policy Development and Management

All companies should develop and maintain clear and robust policies for safeguarding critical business data and sensitive information, protecting their reputation and discouraging inappropriate behavior by employees.

Many of these types of policies already exist for "real world" situations, but may need to be tailored to your organization and updated to reflect the increasing impact of cyberspace on everyday transactions, both professional and personal. As with any other business document, cyber security policies should follow good design and governance practices -- not so long that they become unusable, not so vague that they become meaningless, and reviewed on a regular basis to ensure that they stay pertinent as your business needs change.

Please note that this document does not address all policy requirements for businesses that fall under legislative acts or directives, such as the Health Insurance Portability and Accountability Act, Sarbanes-Oxley Act or other federal, state or local statutes.

Cyber Plan Action Items:

1. Establish security roles and responsibilities

One of the most effective and least expensive means of preventing serious cyber security incidents is to establish a policy that clearly defines the separation of roles and responsibilities with regard to systems and the information they contain. Many systems are designed to provide for strong Role-Based Access Control (RBAC), but this tool is of little use without well-defined procedures and policies to govern the assignment of roles and their associated constraints. Such policies need to clearly state, at a minimum:

- Clearly identify company data ownership and employee roles for security oversight and their inherit privileges, including:
    - Necessary roles, and the privileges and constraints accorded to those roles.
    - The types of employees who should be allowed to assume the various roles.
    - How long an employee may hold a role before access rights must be reviewed.
    - If employees may hold multiple roles, the circumstances defining when to adopt one role over another.

Depending on the types of data regularly handled by your business, it may also make sense to create separate policies governing who is responsible for certain types of data. For example, a business that handles large volumes of personally identifiable information (PII) from its customers may benefit from identifying a chief steward for customers' privacy information. The steward could serve not only as a subject matter expert on all matters of privacy, but also to serve as the champion for process and technical improvements to PII handling.

2. Establish an employee Internet usage policy

The limits on employee Internet usage in the workplace vary widely from business to business. Your guidelines should allow employees the maximum degree of freedom they require to be productive (short breaks to surf the web or perform personal tasks online have been shown to increase productivity). At the same time, rules of behavior are necessary to ensure that all employees are aware of boundaries, both to keep them safe and to keep your company successful. Some to consider:

- Personal breaks to surf the web should be limited to a reasonable amount of time and to certain types of activities.
- If you use a web filtering system, employees should have clear knowledge of how and why their web activities will be monitored, and what types of sites are deemed unacceptable by your policy.
- Workplace rules of behavior should be clear, concise and easy to follow. Employees should feel comfortable performing both personal and professional tasks online without making judgment calls as to what may or may

not be deemed appropriate. Businesses may want to include a splash warning upon network sign-on that advises the employees of the businesses' Internet usage policies so that all employees are on notice.

3. Establish a social media policy

Social networking applications present a number of risks that are difficult to address using technical or procedural solutions. A strong social media policy is crucial for any business that seeks to use social networking to promote its activities and communicate with its customers. At a minimum, a social media policy should clearly include the following:

- Specific guidance on when to disclose company activities using social media, and what kinds of details can be discussed in a public forum.
- Additional rules of behavior for employees using personal social networking accounts to make clear what kinds of discussion topics or posts could cause risk for the company.
- Guidance on the acceptability of using a company email address to register for, or get notices from, social media sites.
- Guidance on selecting long and strong passwords for social networking accounts, since very few social media sites enforce strong authentication policies for users.

Lastly, all users of social media need to be aware of the risks associated with social networking tools and the types of data that can be automatically disclosed online when using social media. Taking the time to educate your employees on the potential pitfalls of social media use, especially in tandem with geo-location services, may be the most beneficial social networking security practice of all.

4. Identify potential reputation risks

All organizations should take the time to identify potential risks to their reputation and develop a strategy to mitigate those risks via policies or other measures as available. Specific types of reputation risks include:

- Being impersonated online by a criminal organization (e.g., an illegitimate website spoofing your business name and copying your site design, then attempting to defraud potential customers via phishing scams or other method).
- Having sensitive company or customer information leaked to the public via the web.
- Having sensitive or inappropriate employee actions made public via the web or social media sites.

All businesses should set a policy for managing these types of risks and plans to address such incidents if and when they occur. Such a policy should cover a regular process for identifying potential risks to the company's reputation in cyberspace, practical measures to prevent those risks from materializing and reference plans to respond and recover from potential incidents as soon as they occur.

Helpful links

- US-CERT's Protect Your Workplace Posters & Brochure:
  http://www.us-cert.gov/reading_room/distributable.html
- Socializing Securely: Using Social Networking Services:
  http://www.us-cert.gov/reading_room/safe_social_networking.pdf
- Governing for Enterprise Security:
  http://www.cert.org/governance/
- FFIEC Handbook Definition of Reputation Risk:
  http://ithandbook.ffiec.gov/it-booklets/retail-payment-systems/retail-payment-systems-risk-management/reputation-risk.aspx

- What Businesses can do to help with cyber security;
  http://www.staysafeonline.org/sites/default/files/resource_documents/What%20Businesses%20Can%20Do%202011%20Final_0.pdf

Cyber Security Glossary

Adware

Any software application that displays advertising banners while the program is running. Adware often includes code that tracks a user's personal information and passes it on to third parties without the user's authorization or knowledge. And if you gather enough of it, adware slows down your computer significantly. Over time, performance can be so degraded that you may have trouble working productively. See also **Spyware** and **Malware**.

Anti-Virus Software

Software designed to detect and potentially eliminate viruses before they have had a chance to wreak havoc within the system. Anti-virus software can also repair or quarantine files that have already been infected by virus activity. See also **Virus** and **Electronic Infections**.

Application

Software that performs automated functions for a user, such as word processing, spreadsheets, graphics, presentations and databases—as opposed to operating system (OS) software.

Attachment

A file that has been added to an email—often an image or document. It could be something useful to you or something harmful to your computer. See also **Virus**.

Authentication

Confirming the correctness of the claimed identity of an individual user, machine, software component or any other entity.

Authorization

The approval, permission or empowerment for someone or something to do something.

Backdoor

Hidden software or hardware mechanism used to circumvent security controls.

Backup

File copies that are saved as protection against loss, damage or unavailability of the primary data. Saving methods include high-capacity tape, separate disk sub-systems or on the Internet. Off-site backup storage is ideal, sufficiently far away to reduce the risk of environmental damage such as flood, which might destroy both the primary and the backup if kept nearby.

Badware

See **Malware**, **Adware** and **Spyware**.

Bandwidth

> The capacity of a communication channel to pass data such as text, images, video or sound through the channel in a given amount of time. Usually expressed in bits per second.

Blacklisting Software

> A form of filtering that blocks only websites specified as harmful. Parents and employers sometimes use such software to prevent children and employees from visiting certain websites. You can add and remove sites from the "not permitted" list. This method of filtering allows for more full use of the Internet, but is less efficient at preventing access to any harmful material that is not on the list. See also **Whitelisting Software**.

Blended Threat

> A computer network attack that seeks to maximize the severity of damage and speed of contagion by combining methods—for example, using characteristics of both viruses and worms. See also **Electronic Infection**.

Blog

> Short for "Web log," a blog is usually defined as an online diary or journal. It is usually updated frequently and offered in a dated log format with the most recent entry at the top of the page. It often contains links to other websites along with commentary about those sites or specific subjects, such as politics, news, pop culture or computers.

Broadband

> General term used to refer to high-speed network connections such as cable modem and Digital Subscriber Line (DSL). These types of "always on" Internet connections are actually more susceptible to some security threats than computers that access the Web via dial-up service.

Browser

> A client software program that can retrieve and display information from servers on the World Wide Web. Often known as a "Web browser" or "Internet browser," Examples include Microsoft's Internet Explorer, Google's Chrome, Apple's Safari, and Mozilla's Firefox.

Brute Force Attack

> An exhaustive password-cracking procedure that tries all possibilities, one by one. See also **Dictionary Attack** and **Hybrid Attack**.

Clear Desk Policy

> A policy that directs all personnel to clear their desks at the end of each working day, and file everything appropriately. Desks should be cleared of all documents and papers, including the contents of the "in" and "out" trays —not simply for cleanliness, but also to ensure that sensitive papers and documents are not exposed to unauthorized persons outside of working hours.

Clear Screen Policy

A policy that directs all computer users to ensure that the contents of the screen are protected from prying eyes and opportunistic breaches of confidentially. Typically, the easiest means of compliance is to use a screen saver that engages either on request or after a specified short period of time. See also **Shoulder Surfing**.

Cookie

A small file that is downloaded by some websites to store a packet of information on your browser. Companies and organizations use cookies to remember your login or registration identification, site preferences, pages viewed and online "shopping-cart" so that the next time you visit a site, your stored information can automatically be pulled up for you. A cookie is obviously convenient but also presents potential security issues. You can configure your browser to alert you whenever a cookie is being sent. You can refuse to accept all cookies or erase all cookies saved on your browser.

Credit Card

A card indicating the holder has been granted a line of credit. Often sought after by criminals looking for an easy way to purchase things without having to pay for them. For this reason and others, a credit card preferable to a debit card for online shopping since it provides a buffer between buyer and seller, affording more protections to the buyer in case there is a problem with the order or the card number is compromised. See also **Debit Card**.

Cyberbullying

Sending or posting harmful, cruel, rude or threatening messages, or slanderous information, text or images using the Internet or other digital communication devices.

Debit Card

A card linked directly to the holder's bank account, withdrawing money from the account. Not as safe as credit cards for online shopping since if problems arise, the buyer's money has already been spent and is harder to get back. See also **Credit Card**.

Denial of Service Attack

The prevention of authorized access to a system resource or the delaying of system operations and functions. Often this involves a cyber criminal generating a large volume of data requests. See also **Flooding**.

Dictionary Attack

A password-cracking attack that tries all of the phrases or words in a dictionary. See also **Brute Force Attack** and **Hybrid Attack**.

Digital Certificate

The electronic equivalent of an ID card that establishes your credentials when doing business or other transactions on the Web. It contains your name, a serial number, expiration dates, a copy of the certificate holder's public key (used for encrypting messages and digital signatures) and the digital signature of the certificate-issuing authority so that a recipient can verify that the certificate is real.

Domain Hijacking

An attack in which an attacker takes over a domain by first blocking access to the domain's DNS server and then putting his own server up in its place.

Domain Name System (DNS)

The DNS is the way that Internet domain names are located. A website's domain name is easier to remember than its IP (Internet Protocol) address.

Dumpster Diving

Recovering files, letters, memos, photographs, IDs, passwords, checks, account statements, credit card offers and more from garbage cans and recycling bins. This information can then be used to commit identity theft.

Electronic Infections

Often called "viruses," these malicious programs and codes harm your computer and compromise your privacy. In addition to the traditional viruses, other common types include worms and Trojan horses. They sometimes work in tandem to do maximum damage. See also Blended Threat.

Encryption

A data security technique used to protect information from unauthorized inspection or alteration. Information is encoded so that it appears as a meaningless string of letters and symbols during delivery or transmission. Upon receipt, the information is decoded using an encryption key.

End User License Agreement (EULA)

A contract between you and your software's vendor or developer. Many times, the EULA is presented as a dialog box that appears the first time you open the software and forces you to check "I accept" before you can proceed. Before accepting, though, read through it and make sure you understand and are comfortable with the terms of the agreement. If the software's EULA is hard to understand or you can't find it, beware!

Evil Twins

A fake wireless Internet hot spot that looks like a legitimate service. When victims connect to the wireless network, a hacker can launch a spying attack on their transactions on the Internet, or just ask for credit card information in the standard pay-for-access deal. See also **Man-in-the-Middle Attacks**.

File-Sharing Programs

Sometimes called peer-to-peer (P2P) programs, these allow many different users to access the same file at the same time. These programs are often used to illegally upload and download music and other software. Examples include Napster, Grokster, Kazaa, iMesh, Ares and Limewire.

Firewall

A hardware or software link in a network that inspects all data packets coming and going from a computer, permitting only those that are authorized to reach the other side.

Flooding

An attack that attempts to cause a failure in the security of a computer by providing more input, such as a large volume of data requests, than it can properly process. See also **Denial of Service Attack**.

Grooming

Using the Internet to manipulate and gain trust of a minor as a first step towards the future sexual abuse, production or exposure of that minor. Sometimes involves developing the child's sexual awareness and may take days, weeks, months or in some cases years to manipulate the minor.

Hacker

An individual who attempts to break into a computer without authorization.

HTTPS

When used in the first part of a URL (e.g., http://), this term specifies the use of hypertext transfer protocol (HTTP) enhanced by a security mechanism such as Secure Socket Layer (SSL). Always look for the HTTP*S* on the checkout or order form page when shopping online or when logging into a site and providing your username and password.

Hybrid Attack

Builds on other password-cracking attacks by adding numerals and symbols to dictionary words. See also **Dictionary Attack** and **Brute Force Attack**.

Instant Messaging (IM)

A service that allows people to send and get messages almost instantly. To send messages using instant messaging you need to download an instant messaging program and know the instant messaging address of another person who uses the same IM program. See also **Spim**.

IP (Internet Protocol) Address

A computer's inter-network address, written as a series of four 8-bit numbers separated by periods, such as 123.45.678.990. Every website has an IP Address, although finding a website is considerably easier to do when using its domain name instead. See also **Domain Name System (DNS)**.

Internet Service Provider (ISP)

A company that provides internet access to customers.

Keystroke Logger

A specific type of electronic infection that records victims' keystrokes and sends them to an attacker. This can be done with either hardware or software. See also **Trojan Horse**.

Malware

A generic term for a number of different types of malicious code. See also **Adware** and **Spyware**.

Man-In-the-Middle Attack

Posing as an online bank or merchant, a cyber criminal allows a victim to sign in over a Secure Sockets Layer (SSL) connection. The attacker then logs onto the real server using the client's information and steals credit card numbers.

Monitoring Software

Software products that allow parents to monitor or track the websites or email messages that a child visits or reads. See also **Blacklisting Software** and **Whitelisting Software**.

Network

Two or more computer systems that are grouped together to share information, software and hardware.

Operating System (OS)

Programs that manage all the basic functions and programs on a computer, such as allocating system resources, providing access and security controls, maintaining file systems and managing communications between end users and hardware devices. Examples include Microsoft's Windows, Apple's Macintosh and Red Hat's Linux.

Password

A secret sequence of characters that is used as a means of authentication to confirm your identity in a computer program or online.

Password Cracking

Password cracking is the process of attempting to guess passwords, given the password file information. See also **Brute Force Attacks**, **Dictionary Attacks** and **Hybrid Attacks**.

Password Sniffing

Passive wiretapping, usually on a local area network, to gain knowledge of passwords.

Patch

A patch is a small security update released by a software manufacturer to fix bugs in existing programs. Your computer's software programs and/or operating system may be configured to check automatically for patches, or you may need to periodically visit the manufacturers' websites to see if there have been any updates.

Peer-to-Peer (P2P) Programs

See **File-Sharing Programs**.

Phishing

Soliciting private information from customers or members of a business, bank or other organization in an attempt to fool them into divulging confidential personal and financial information. People are lured into sharing user names, passwords, account information or credit card numbers, usually by an official-looking message in an email or a pop-up advertisement that urges them to act immediately, usually by clicking on a link provided. See also **Vishing**.

Pharming

Redirecting visitors from a real website to a bogus one. A user enters what is believed to be a valid Web address and is unknowingly redirected to an illegitimate site that steals the user's personal information. On the spoofed site, criminals may mimic real transactions and harvest private information unknowingly shared by users. With this, the attacker can then access the real website and conduct transactions using the credentials of a valid user.

Router

A hardware device that connects two or more networks and routes incoming data packets to the appropriate network. Many Internet Service Providers (ISPs) provide these devices to their customers, and they often contain firewall protections.

Script

A file containing active content -- for example, commands or instructions to be executed by the computer.

Shoulder Surfing

Looking over a person's shoulder to get confidential information. It is an effective way to get information in crowded places because it's relatively easy to stand next to someone and watch as they fill out a form, enter a PIN number at an ATM machine or type a password. Can also be done long-distance with the aid of binoculars or other vision-enhancing devices. To combat it, experts recommend that you shield paperwork or your keypad from view by using your body or cupping your hand. Also, be sure you password-protect your computer screen when you must leave it unattended, and clear your desk at the end of the day. See also **Clear Desk Policy** and **Clear Screen Policy**.

Skimming

A high-tech method by which thieves capture your personal or account information from your credit card, driver's license or even passport using an electronic device called a "skimmer." Such devices can be purchased online for under $50. Your card is swiped through the skimmer and the information contained in the magnetic strip on the card is then read into and stored on the device or an attached computer. Skimming is predominantly a tactic used to perpetuate credit card fraud, but is also gaining in popularity amongst identity thieves.

Social Engineering

A euphemism for non-technical or low-technology means—such as lies, impersonation, tricks, bribes, blackmail and threats—used to attack information systems. Sometimes telemarketers or unethical employees employ such tactics.

Social Networking Websites

Sites specifically focused on the building and verifying of social networks for whatever purpose. Many social networking services are also blog hosting services. There are more than 300 known social networking websites, including Facebook, MySpace, Friendster, Xanga and Blogspot. Such sites enable users to create online profiles and post pictures and share personal data such as their contact information, hobbies, activities and interests. The sites facilitate connecting with other users with similar interests, activities and locations. Sites vary in who may view a user's profile—some have settings which may be changed so that profiles can be viewed only by "friends." See also **Blogs**.

Spam

Unwanted, unsolicited email from someone you don't know. Often sent in an attempt to sell you something or get you to reveal personal information.

Spim

Unwanted, unsolicited instant messages from someone you don't know. Often sent in an attempt to sell you something or get you to reveal personal information.

Spoofing

Masquerading so that a trusted IP address is used instead of the true IP address. A technique used by hackers as a means of gaining access to a computer system.

Spyware

Software that uses your Internet connection to send personally identifiable information about you to a collecting device on the Internet. It is often packaged with software that you download voluntarily, so that even if you remove the downloaded program later, the spyware may remain. See also **Adware** and **Malware**.

SSL (Secure Socket Layer)

An encryption system that protects the privacy of data exchanged by a website and the individual user. Used by websites whose URLs begin with https instead of http.

Trojan Horse

A computer program that appears to be beneficial or innocuous, but also has a hidden and potentially malicious function that evades security mechanisms. A "keystroke logger," which records victims' keystrokes and sends them to an attacker, or remote-controlled "zombie computers" are examples of the damage that can be done by Trojan horses. See also **Electronic Infection**.

URL

Abbreviation for "Uniform (or Universal) Resource Locator." A way of specifying the location of publicly available information on the Internet. Also known as a Web address.

URL Obfuscation

Taking advantage of human error, some scammers use phishing emails to guide recipients to fraudulent sites with names very similar to established sites. They use a slight misspelling or other subtle difference in the URL, such as "monneybank.com" instead of "moneybank.com" to redirect users to share their personal information unknowingly.

Virus

A hidden, self-replicating section of computer software, usually malicious logic, that propagates by infecting—i.e., inserting a copy of itself into and becoming part of -- another program. A virus cannot run by itself; it requires that its host program be run to make the virus active. Often sent through email attachments. Also see **Electronic Infection** and **Blended Threat**.

Vishing

Soliciting private information from customers or members of a business, bank or other organization in an attempt to fool them into divulging confidential personal and financial information. People are lured into sharing user names, passwords, account information or credit card numbers, usually by an official-looking message in an email or a pop-up advertisement that urges them to act immediately—but in a vishing scam, they are urged to call the phone number provided rather than clicking on a link. See also **Phishing**.

Vulnerability

A flaw that allows someone to operate a computer system with authorization levels in excess of that which the system owner specifically granted.

Whitelisting Software

A form of filtering that only allows connections to a pre-approved list of sites that are considered useful and appropriate for children. Parents sometimes use such software to prevent children from visiting all but certain websites. You can add and remove sites from the "permitted" list. This method is extremely safe, but allows for only extremely limited use of the Internet.

Worm

Originally an acronym for "Write once, read many times," a type of electronic infection that can run independently, can propagate a complete working version of itself onto other hosts on a network, and may consume computer resources destructively. Once this malicious software is on a computer, it scans the network for another machine with a specific security vulnerability. When it finds one, it exploits the weakness to copy itself to the new machine, and then the worm starts replicating from there, as well. See also **Electronic Infection** and **Blended Threat**.

Zombie Computer

A remote-access Trojan horse installs hidden code that allows your computer to be controlled remotely. Digital thieves then use robot networks of thousands of zombie computers to carry out attacks on other people and cover up their tracks. Authorities have a harder time tracing criminals when they go through zombie computers.

Sources:

National Institute of Standards and Technology:
http://csrc.nist.gov/publications/nistir/ir7298-rev1/nistir-7298-revision1.pdf

Whoiswatchingcharlottesville.org:
http://www.whoswatchingcharlottesville.org/glossary.html

Cyber Security Links

Cyber Security and Privacy Protection

- Carnegie Mellon Software Engineering Institute's CERT Coordination Center:
  www.cert.org/other_sources

- Center for Internet Security (CIS):
  www.cisecurity.org

- Free online security check ups:
  http://www.staysafeonline.org/tools-resources/free-security-check-ups

- National Cyber Security Alliance for Small Business Home Users:
  http://www.staysafeonline.info/

- OnGuard Online:
  www.OnGuardOnline.gov

- SANS (SysAdmin, Audit, Network, Security) Institute's Most Critical Internet Security Vulnerabilities:
  www.sans.org/top20

- Security Tips from Securing our eCity:
  http://securingourecity.org/

- Small Business Solutions form StopBadware:
  http://stopbadware.org/

- The Open Web Application Security Project:
  www.owasp.org

Cyber Security Threat Centers

- McAfee Cybersafety Resource Portal
  http://www.mcafee.com/cru

- McAfee Security Solutions for Small Business:
  http://shop.mcafee.com/Default.aspx?site=us&pid=HOME&CID=MFE-MHP001

- Symantec Security Solutions for Small Business:
  http://store.symantec.com/?om_sem_cid=hho_sem_nam_us_Google_SMB_Store_Home&inid=hho_sem_s
  y:us:ggl:en:e%7Ckw0000006084%7CSMB

Training and Exercises

- Free training materials, security configuration guides from Internet Security Alliance:
  http://www.isalliance.org/

- NIH Free Online User Training:
  http://iase.disa.mil/eta/issv4/index.htm

- NIH Free Online User Training (non DOD version):
  http://irtsectraining.nih.gov/publicUser.aspx

Government Resources

- Department of Homeland Security (DHS)'s National Strategy to Secure Cyberspace:
  www.dhs.gov/xlibrary/assets/National_Cyberspace_Strategy.pdf

- DHS testimony before the House on Committee on Homeland Security Subcommittee on Cybersecurity, Infrastructure Protection, and Security Technologies:
  http://www.dhs.gov/ynews/testimony/testimony_1300283858976.shtm

- FCC Cyber Security Encyclopedia Page
  http://www.fcc.gov/cyberforsmallbiz

- FCC Public Safety and Homeland Secuirity Bureau Clearinghouse:
  http://publicsafety.fcc.gov/pshs/clearinghouse/index.htm

- FCC Public Safety and Homeland Security Bureau Guidelines for Emergency Planning:
  http://transition.fcc.gov/pshs/emergency-information/guidelines/

- FCC Ten Cybersecurity Tips for Small Businesses
  http://hraunfoss.fcc.gov/edocs_public/attachmatch/DOC-306595A1.pdf

- Federal Trade Commission Guide for Business
  http://www.ftc.gov/bcp/edu/microsites/infosecurity/

- Federal Trade Commission – Identity Theft Information:
  http://www.onguardonline.gov/topics/computer-security.aspx

- Federal Trade Commission's Interactive Tutorial:
  www.ftc.gov/infosecurity

- National Institute of Standards and Technology (NIST)'s Computer Security Resource Center:
  www.csrc.nist.gov

- NIST briefing on Cybersecurity for Small Businesses:
  http://csrc.nist.gov/groups/SMA/sbc/documents/smb-presentation.pdf

Government Resources (cont'd)

- NIST Guide to Selecting Information Technology Security Products:
  http://csrc.nist.gov/publications/nistpubs/800-36/NIST-SP800-36.pdf

- NIST's Risk Management Guide for Information Technology Systems:
  www.csrc.nist.gov/publications/nistpubs/800-30/sp800-30.pdf

- NIST Small Business Corner - A link to the NIST-SBA-FBI Small Business Information Security outreach pages :
  http://csrc.nist.gov/groups/SMA/sbc/index.html

- NIST Small Business Information Security:
  http://csrc.nist.gov/publications/nistir/ir7621/nistir-7621.pdf

- SBA, NIST and FBI partnership on Cybersecurity for small businesses:
  http://csrc.nist.gov/groups/SMA/sbc/overview.html

- United States Computer Emergency Readiness Team (US-CERT):
  www.us-cert.gov

- U.S. Department of Homeland Security Cyber Security Resources:
  http://www.dhs.gov/cyber

Publications

- 2011 Awards for best computer security tools, SC Magazine:
  http://www.scmagazineus.com/2011-sc-awards-us-finalists/section/1908/

- Cloud Security Alliance
  https://cloudsecurityalliance.org/csaguide.pdf

- Computer Security Resource Center, National Instiitute of Standards and Technology:
  http://csrc.nist.gov/groups/SMA/sbc/library.html

- Microsoft Small Business Guide:
  http://www.microsoft.com/smallbusiness/support/security-toolkit-pdf.mspx

- Protecting Your Small Business, Entrepreneur Magazine:
  http://www.entrepreneur.com/magazine/entrepreneur/2010/june/206656.html

- Small business Information Security: The Fundamentals, National Institute of Standards and Technology:
  http://csrc.nist.gov/publications/nistir/ir7621/nistir-7621.pdf

www.ingramcontent.com/pod-product-compliance
Lightning Source LLC
Chambersburg PA
CBHW081233170526
45165CB00009B/3048